TAXONOMY TRIVIA

WRITTEN BY
THE GOOD AND THE BEAUTIFUL TEAM

DESIGNED BY
ANNA ASFOUR

COVER DESIGN BY **ROBIN FIGHT**

© 2022 JENNY PHILLIPS | goodandbeautiful.com

What is the mountain lion most closely related to?

A. TIGER   B. HOUSE CAT   C. DOG   D. LION

A mountain lion, otherwise known as a cougar, is not really a lion at all, but instead is more closely related to the HOUSE CAT. Mountain lions (cougars) purr and meow like other small cats, even though they are deadly hunters.

As a meat eater, a raven is in the same order as a hawk.

A. TRUE   B. FALSE

FALSE. The raven, though it mostly eats meat, is actually the largest member of the Passeriformes order, or perching birds. With a wingspan of over 1.3 m (4 ft), the raven is an intelligent bird, able to kill animals as large as young sheep. Ravens have even learned how to lead wolves to their prey, giving these birds the advantage of knowing where to scavenge what is left of the kill later on. These beautiful birds can make a variety of vocalizations.

Which of these animals is a close living relative to the elephant?

A. MANATEE   B. GORILLA

MANATEES are closely related to elephants! Both manatees and elephants are part of a group of animals called Uranotherians.

Some fish can live out of water.

A. TRUE   B. FALSE

TRUE. A lungfish breathes with both gills and an organ similar to lungs that allows it to breathe air. The South American and African lungfishes burrow into the muddy bottoms of lakes and rivers. If the body of water dries up, the lungfishes can survive completely out of water for months, or even years, at a time.

Bats are part of the rodent family.

A. TRUE   B. FALSE

FALSE. Bats are not even closely related to rodents such as mice and rats. Bats are part of the order Chiroptera, which means "hand-wing" in Greek. The order Chiroptera includes more than 1,200 species of bats! Bats are also the only mammal that can fly.

Which animal is the red panda's closest relative?

A. CAT   B. GIANT PANDA   C. SKUNK   D. MONKEY

Though its name may lead you to believe the red panda is a giant panda relative, it is actually more closely related to a SKUNK. The red panda is the only living organism in the family Ailuridae and is an ancient species from the order Carnivora, although its diet consists mainly of bamboo.

Which big cat likes water best?

A. LEOPARD   B. LION   C. JAGUAR   D. TIGER

TIGERS often live around wet habitats such as tropical rainforests, swamps, and wetlands. They are excellent swimmers who will often dive into the water with great enthusiasm. Sometimes tigers find food to eat in the water; however, these playful cats may just plunge into the refreshing water to keep cool on a sultry day.

What is the fastest animal on Earth?

A. PEREGRINE FALCON   B. CHEETAH   C. HUMMINGBIRD   D. SAILFISH

When this incredible bird swoops down to grab its prey, you would not want to be anywhere close by. With powerful wings and a sleek body, it can dive through the air with a speed topping 300 km (186 mi) per hour! With the cheetah clocking just 114 km (71 mi) per hour, the PEREGRINE FALCON wins the race for speed by a landslide.

Which of these animals is the closest living relative to the tapir?

A. COW   B. ANTEATER   C. PIG   D. RHINOCEROS

Tapirs may look like anteaters with their long snouts, but they are actually most closely related to RHINOCEROSES and horses. This is because of their odd number of toes, with three on each back foot and four on each front foot.

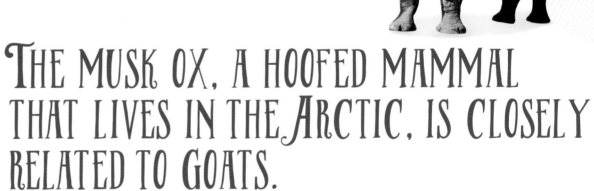

TAPIR

The musk ox, a hoofed mammal that lives in the Arctic, is closely related to goats.

A. TRUE   B. FALSE

MUSK OX

TRUE. Musk oxen are members of the Bovidae family and are ungulates; they are named for their strong musky odor. This family also includes chamois [sha-MWAH] and mountain goats, making musk oxen closely related to domestic goats and sheep.

A GIRAFFE HAS THE SAME NUMBER OF NECK BONES AS A HUMAN.

A. TRUE   B. FALSE

TRUE. Surprisingly, a giraffe's neck has seven bones which is the same number of vertebrae in a human neck. The giraffe's neck bones grow to around 25 cm (10 in) long. In order to pump blood up its long neck, this amazing creature also boasts an enormous heart, which weighs 11 kg (24.2 lb) and is 0.6 m (2 ft) long. No wonder giraffes have the highest blood pressure of any animal in the world!

What is the largest marine animal that eats plants?

A. DOLPHIN   B. BLUE WHALE

C. MANATEE   D. SEAL

MANATEES are constantly eating while awake. Average-sized manatees tip the scales at around 544 kg (1,200 lb)! They devour around 10% of their body weight in grass each day, making them the largest vegetarians in the sea. Due to their vast consumption of food, manatees continue growing even as adults.

OTTER

An organism in the order Carnivora would most likely eat _____.

A. LEAVES   B. FLESH   C. SEEDS   D. APPLES

Organisms in the order Carnivora specialize in eating **FLESH**. These members are more commonly referred to as carnivores. Some species in this order may be omnivores or herbivores, such as raccoons or giant pandas.

Which organism belongs to the kingdom Monera?

A. TOMATO   B. MUSHROOM   C. GRASSHOPPER   D. E. COLI

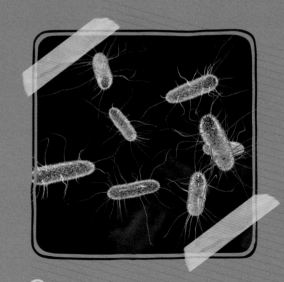

The kingdom Monera is made up of prokaryote organisms such as bacteria. These are single-celled organisms that lack a true nucleus. Escherichia coli **(E. COLI)** is a bacterium that normally resides in the human body. However, ingesting food or water that is contaminated with E. coli can cause digestive upset.

The American pronghorn antelope is most closely related to the giraffe.

A. TRUE   B. FALSE

TRUE. The American pronghorn antelope is not actually an antelope at all. Unlike true antelope, such as those belonging to the family Bovidae (gazelles and impalas), the American pronghorn antelope is not closely related to sheep and cattle. Rather, it is the only living species in the family Antilocapridae, making its closest living relative the giraffe.

How many living species of monotremes (egg-laying mammals) exist?

A. NONE   B. ONE   C. FIVE   D. MORE THAN 4,000

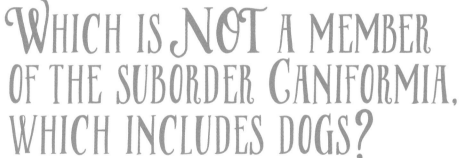

FIVE. The duck-billed platypus, often described as a mix of a duck, beaver, and otter, is a monotreme. There are also four types of spiny echidnas, more commonly known as spiny anteaters, that are egg-laying mammals. Strangely, all monotremes are now only found in Australia and New Guinea.

Which is NOT a member of the suborder Caniformia, which includes dogs?

A. SEALS   B. CATS   C. BEARS   D. RACCOONS

CATS! Yes, seals are related to dogs! Caniformia includes dogs, seals, and other "dog-like" carnivores. Both dogs and seals eat meat—some seals even eat penguins! Even though they are in the same suborder, seals are in the family Pinnipeds, which includes other species of fin-footed mammals such as the walrus. Dogs are in the family Canidae.

Roses are in the same family as apples, plums, and almonds.

A. TRUE   B. FALSE

TRUE. The plant family Rosaceae is made of thousands of plant species. Among those are roses, apples, plums, and raspberries. Characteristics of this family include woody stems (often with thorns), simple or compound leaves, large flowers with five petals, and stipules at the base of the leaf.

The world's smallest flowering plant has leaves and roots.

A. TRUE   B. FALSE

FALSE. The world's smallest flowering plant is the Wolffia arrhiza, commonly known as "watermeal," and is native to Africa, Europe, and some parts of Asia. These tiny plants measure only about 1 mm (0.04 in)! You won't find all of the normal plant parts on this little wonder. It doesn't have leaves, stems, or roots. If conditions are ideal, it will sometimes grow a little flower with one stamen and one pistil.

Which unusual tree is named after a fictional animal but looks like an umbrella?

A. ELEPHANT EAR   B. DRAGON'S BLOOD TREE

C. DOGWOOD TREE   D. CATNIP

The DRAGON'S BLOOD TREE is only found on the island of Socotra in the Indian Ocean. The tree is named for its sap, known as "dragon's blood," which is a deep red color even after it has dried into resin. Its shape looks very much like an umbrella, with a densely packed canopy that acts as a shade and prevents evaporation of the little amount of rain it receives every year.

Which of these little plants in the Aizoaceae family survives by pretending to be a rock?

A. FAIRY ELEPHANT'S FEET   B. CARPET OF STARS

C. LITHOPS   D. CHILEAN SEA FIG

LITHOPS are small two-leafed plants native to southern Africa and are commonly called "living stones." They grow to only 2.5–5 cm (1–2 in) wide and stay flush to the ground. They can be incredibly difficult to find in their natural habitat because they look like small rocks. There are at least 37 species of this unique little plant that come in a variety of colors, including brown, pink, orange, green, and gray. Many have a small white, yellow, or pale orange daisy-like flower that blooms during early winter.

Most traditional tea varieties all come from the same plant.

A. TRUE   B. FALSE

TRUE. A single plant, Camellia sinensis (or tea plant), is used to make black tea, green tea, oolong tea, and white tea. Different processing methods, such as drying, oxidation, and aging, are used to make different types of tea from the same plant.

The Nepenthes attenboroughii is a carnivorous plant that eats _____.

A. BEETLES   B. SMALL RODENTS   C. INSECTS   D. ALL OF THE ABOVE

ALL OF THE ABOVE! The Nepenthes attenboroughii is the largest known carnivorous plant in the world, growing to heights of 1.5 m (4.9 ft) tall. The pitcher part of this plant can reach 30 cm (11.8 in) in diameter and is capable of devouring animals from the size of a beetle to small rodents, such as mice and shrews. The plant was only recently discovered on Mount Victoria in the Philippines. Isn't it amazing that we are still discovering new things on God's beautiful creation?

Rafflesia arnoldii, the largest individual flower in the world, has a bloom that measures _____.

A. 15 CM (6 IN)   B. 30.5 CM (1 FT)   C. MORE THAN 1 M (3.3 FT)

The largest recorded individual flower in the world is the Rafflesia arnoldii which grows **MORE THAN 1 M (3.3 FT)** in diameter! Curiously, this flower has no leaves, roots, or stems. It is considered a parasitic plant that only grows on a specific type of vine. It is also considered one of the worst-smelling flowers in the world.

The largest living organism on Earth is a _____.

A. TREE   B. ANIMAL   C. FUNGUS   D. FLOWER

As unbelievable as it sounds, the largest living organism on Earth is a **FUNGUS** called Armillaria ostoyae, or honey mushroom. It grows in the Malheur National Forest in Oregon and covers more than 9.7 sq km (3.7 sq mi) and is still growing! Most of it grows completely underground as flat, stringy structures called rhizomorphs. Scientists are able to track its growth by following the areas of dead and dying trees, which the parasitic organism feeds on. The only other indication of its presence can be found in the autumn season by looking for the clumps of small yellow-brown mushrooms that grow above ground for a short period.

Animal classification is continuously changing.

A. TRUE   B. FALSE

TRUE. Scientists use a number of methods to identify and classify animals, plants, and organisms. Many of those methods rely on equipment that is dependent on advancing technologies. As equipment and technology become more sophisticated, what we know about living things gets more detailed and we discover new things we didn't know before! Sometimes new species are found, and sometimes the classification of known species can change.

This animal is commonly called a bear, but it's really a marsupial. What is it?

A. SLOTH BEAR   B. KODIAK BEAR   C. KOALA BEAR   D. SPECTACLED BEAR

Though it's often referred to as a KOALA BEAR, this animal is actually a member of the marsupial family, which also includes kangaroos, opossums, wombats, and wallabies. Koalas are native to Australia, where they inhabit open eucalyptus woodlands.

The penguin is the only bird that can swim underwater.

A. TRUE   B. FALSE

FALSE. There are actually a few birds that can swim underwater! Two outstanding examples are the American dipper and the cormorant. American dippers can not only "fly" underwater, but they can also walk on the bottoms of swiftly moving streams! They are also North America's only aquatic songbird. The cormorant will float in the water with only part of its neck and bill out of the water. When it spots a tasty meal, it will dive and swim underwater, propelling itself with both its feet and its wings!

Which other toothed whale is most closely related to the narwhal?

A. BELUGA WHALE   B. ORCA (KILLER WHALE)

C. BOTTLENOSE DOLPHIN   D. SPERM WHALE

While BELUGA WHALES and narwhals aren't the same species, they have similarities and have interbred on rare occasions, creating an animal known as a "narluga." Narwhals and belugas make up the entire Monodontidae family.

Flying fish are the only fish that leap out of water.

A. TRUE   B. FALSE

FALSE. Flying fish have the incredible ability to "fly" as far as 200 m (650 ft) when they jump out of water, thanks to specialized fins. They're far from the only jumping fish, though. Many other types of fish will leap out of water at times, including lake trout and even great white sharks. When devil rays leap out of water, they flap their fins and sometimes flip!

The thylacine was a type of dog.

A. TRUE   B. FALSE

THYLACINES

SAND CAT

FALSE. The thylacine [THAI-luh-seen], often referred to as the Tasmanian tiger, looked very much like a dog but was actually a marsupial, part of the Marsupialia infraclass. Once roaming areas of Australia, thylacines became extinct in 1936. They had the incredible ability to open their jaws to nearly a 90° angle and carried their young in pouches like many other marsupials do.

Sand cats are most likely to be found in what type of ecosystem?

A. DESERT   B. BEACH   C. CITY   D. THE TROPICS

Sand cats are similar to house cats in looks and size, but these cats need to live in the ecosystems they were made for: **DESERTS!** Sand cats have special pads on their feet that allow them to walk on very cold and very hot sand. Temperatures where sand cats live can vary from -0.5 °C (31 °F) to 51 °C (124 °F)! Sand cats dig to avoid extreme temperatures, and they love to hunt snakes.

Which of these animals are in the same order as the horse?

A. TAPIR   B. ZEBRA

C. RHINOCEROS   D. ALL OF THE ABOVE

ALL OF THE ABOVE. All of these animals are part of the Perissodactyl mammal order. The animals in this order have either one or three toes on their hind feet, and those toes must be hoofed.

Komodo dragons use their tongues to smell.

A. TRUE

B. FALSE

TRUE. Believe it or not, some animals use something other than their noses in order to smell! In Komodo dragons a specialized tongue does the job. This long tongue forks at the end and is used to gather information from the air. This and another body part called a Jacobson's organ help the Komodo dragon analyze if prey is coming and from which direction. Komodo dragons are members of the Reptilia class of animals. Many other reptiles also use this method to smell.

Hyenas, Caracals, and Mongooses are all part of the Feliformia suborder.

A. TRUE   B. FALSE

TRUE. Suborder Feliformia is a great example of animal classification not being based on looks. Feliformia includes cat-like animals that are part of the Carnivora order, so you probably wouldn't expect to see hyenas or mongooses in this group. Structural makeup is what brings these very different animals into the same suborder. Some animals you may expect to find in Feliformia are caracals, cheetahs, and lions.

Flies have tiny teeth to help them chew their food.

A. TRUE   B. FALSE

FALSE. Flies are part of the Insecta class and do not have teeth to chew their food. Instead, they vomit digestive juices on solid food to turn it into a liquid. Then they use their tongues, which are shaped like drinking straws, to suck up their food.

The baiji was a species of dolphin.

A. TRUE   B. FALSE

TRUE. Have you ever heard of a baiji [BAI-jee]? It's a species of dolphin also known as the Yangtze River dolphin. These dolphins lived in, you guessed it, the Yangtze River in China! This is the first species of dolphin that has been declared extinct due to human activity. Baijis make up the entire Lipotidae mammal family. Check out their tiny eyes!

BAIJI

The Lagomorpha order is composed of only rabbits and hares.

A. TRUE   B. FALSE

HARE

PIKA

FALSE. Less well known than the other members of its order, the pika looks quite different than a rabbit or a hare. Pikas have short legs and no tails and look very much like rodents. Hares and rabbits make up the Leporidae family, and pikas belong to the Ochotonidae family. Together these two families create the Lagomorpha order.

A SAIGA IS WHICH TYPE OF ANIMAL?

A. DEER   B. ANTELOPE

SAIGA ANTELOPE

While you may think saiga [SAI-guh] ANTELOPE look like something out of a science fiction story, they are real animals. Due to both disease and poaching, saigas are considered critically endangered. These unique animals live in desert and grassland steppe areas in Kazakhstan, Russia, Uzbekistan, Turkmenistan, and Mongolia. They are the only members of the genus Saiga and are members of the Bovidae family, which includes cattle, goats, and sheep.

HONEY BADGERS HAVE A DEFENSE MECHANISM SIMILAR TO A/AN _____.

A. OTTER   B. FERRET   C. HONEYBEE   D. SKUNK

HONEY BADGER

Like a SKUNK, a honey badger has a gland at the base of its tail that stores a stinky-smelling liquid. Usually, a honey badger uses its powerful stench to mark territory, but a honey badger will also drop a "stink bomb" to ward off predators.

Hippopotamuses live in a habitat that contains diverse species of animals. Which of the following animals is the hippo NOT related to?

A. GIRAFFE B. ANTELOPE
C. ELEPHANT D. WARTHOG

ELEPHANT. Hippos are classified in the order Artiodactyla, more commonly called ungulates, which means that they are hoofed animals with an even number of functional toes. Giraffes, antelope, and warthogs all belong to this same order. The elephant is classified into the Proboscidae order because of its "proboscis," or trunk.

When food is hard to find, marine iguanas _____.

A. SHRINK   B. HIBERNATE   C. BURROW   D. SWIM

Marine iguanas are known as the SHRINKING lizards of the Galápagos Archipelago. They are the only lizards known to search for food in the ocean, eating mostly red and green algae that grow in the more shallow parts of the ocean. When periodic weather events cause the water to warm up, which then causes the red and green algae to disappear, marine iguanas shrink in both weight and length. When the water turns cooler and food becomes more abundant, the marine iguanas will grow back to their original sizes.

Some mushrooms create their own weather.   A. TRUE   B. FALSE

TRUE. Scientists have found that gilled mushrooms, part of the fungal order Agaricales, create their own weather in the form of tiny winds! These gilled mushrooms, which include Shiitake and oyster mushrooms, evaporate water which cools the air under the mushroom's cap. The cooled air then flows away from the mushroom, carrying the spores of the mushroom with it. These tiny air currents can carry the spores up to 10 cm (4 in) horizontally and vertically, and from there they catch a ride on stronger winds to spread far and wide.

The phenomena known as "red tides" are caused by harmful algae blooms.

A. TRUE

B. FALSE

TRUE. The term "red tides" is misleading. Red tides are often but not always red, and they are not related to the ocean tides. They are actually harmful algae blooms caused by a sudden increase in dangerous aquatic microorganisms, such as dinoflagellates and diatoms. The sudden, abundant growth of these protists causes a release of a toxin, or natural poison, that can kill fish and shellfish and also cause the people who eat them to become ill.

Which of these reptiles will squirt blood from its eye when threatened by an enemy?

A. BOA CONSTRICTOR   B. HORNED LIZARD   C. CHAMELEON   D. DESERT TORTOISE

When it feels threatened by an enemy, a HORNED LIZARD can squirt blood from its eye. The blood comes from a duct in the corner of the eye and can travel as far as 1 m (3 ft). This strange defense mechanism doesn't hurt the lizard, but it does confuse predators and contains a chemical that is harmful to dogs, wolves, and coyotes.

Which of these common foods is actually made from bacteria?

A. BANANA   B. YOGURT

C. CHICKEN NUGGET   D. ROMAINE LETTUCE

That cup of YOGURT you ate this morning was actually made with bacteria! Yogurt is made with probiotics (live microorganisms) which can provide protection for your bones and teeth and can help prevent digestive problems. However, many kinds of yogurts have been through a process called pasteurization, which is a heat treatment that kills the beneficial bacteria that were part of the starter culture. To ensure you are receiving the good bacteria found in yogurt, look for products that contain live, active cultures, which should be written on the label.

Potato and tomato don't just rhyme; they're related scientifically!

A. TRUE

B. FALSE

TRUE. Both potato and tomato are members of the Solanaceae, or nightshade, family of flowering plants. This family also contains eggplants, peppers, and other plants that are important in the production of medicines.

Which of the following types of cheese is made from mold?

A. CHEDDAR CHEESE   B. SWISS CHEESE

C. BLEU CHEESE   D. PARMESAN CHEESE

BLEU CHEESE is a type of soft cheese made with cultures of the mold Penicillium. The mold cultures are mixed with the cheese curds, or curdled milk. Bleu cheese is typically stored in a damp, temperature-controlled environment, such as a cave, which helps the mold to grow and gives this cheese its signature spots of blue and green.

Saltwater fish and freshwater fish are classified separately.

A. TRUE

B. FALSE

FALSE. All fish are classified as vertebrates (although hagfish have no vertebrae) and are grouped into classes by their skeletal framework: Agnatha (jawless), Chondrichthyes (cartilaginous), and Osteichthyes (bony). Although fish vary drastically in their habitats and have differing cells that allow them to or prevent them from living in salt water, this is not part of their scientific classification.

FRESHWATER FISH

SALTWATER FISH

Which is NOT a member of the Plantae kingdom?

A. MOSSES   B. FERNS

C. GIANT KELP   D. FLOWERING PLANTS

That's right! Even though it resembles a tall grass, GIANT KELP (and other types of seaweed) is not a plant—it's actually a type of algae, which belongs to the Protista kingdom. When people think of the word algae, they often think of slimy green growth on the surface of ponds and lakes and by the ocean shore. Until fairly recently, many scientists even believed that algaes were plants!

Some types of plants are carnivores.

A. TRUE   B. FALSE

TRUE. When we think of how plants grow, we usually think of soil, water, and sunlight. But, there are some plants that are carnivores, or meat eaters, and they grow by eating insects, crustaceans, and other small animals! Throughout the world there are over 600 species of carnivorous plants, including the well-known Venus flytrap. Carnivorous plants have special leaves and other parts that act as traps and allow them to capture their prey.

Although bears are grouped in the order Carnivora, most species eat more plant matter than meat, especially depending on the season.

A. TRUE B. FALSE

TRUE. All bears eat some plants, but some species of this meat-eating order eat more plant matter than they do meat. It is believed that a black bear's diet contains as much as 85% plants, with some worms, grubs, fish, and small mammals making up the rest.